Weird, Wild, and Wonderful

Marine Minibeasts

Gareth Stevens
Publishing

By Kerry Nagle

Please visit our Web site **www.garethstevens.com.** For a free color catalog of all our high-quality books, call toll free 1-800-542-2595 or fax 1-877-542-2596.

Library of Congress Cataloging-in-Publication Data
Nagle, Kerry.
 Marine minibeasts / Kerry Nagle.
     p. cm.
 Includes index.
 ISBN 978-1-4339-3571-8 (library binding)
 1. Marine invertebrates—Juvenile literature. 2. Marine animals—Juvenile literature. I. Title.
 QL122.2.N24 2010
 592'.177—dc22
                                    2009043879

Published in 2010 by
Gareth Stevens Publishing
111 East 14th Street, Suite 349
New York, NY 10003

© 2010 Blake Publishing

For Gareth Stevens Publishing:
Art Direction: Haley Harasymiw
Editorial Direction: Kerri O'Donnell

Designed in Australia by www.design-ed.com.au

Photography by Kathie Atkinson
Additional photographs: ©iStockphotos.com: Zeynep Mufti, p. 4a, John Anderson, p. 4b, Amanda Rohde, p. 5a, Gergo Orban, p. 5b; © Newspix/News Ltd, p. 19a.

Printed in the United States of America

CPSIA compliance information: Batch #CW10GS: For further information contact Gareth Stevens, New York, New York, at 1-800-542-2595.

Contents

What Are Marine Minibeasts?......... 4

Stuck to One Spot 6

Flowers of the Sea 8

Sea Stars...................................... 10

Spiny Sea Dwellers 12

Sea Slugs 14

Living in a Shell.............................16

Scary Stingers...............................18

Floating Hunters............................ 20

Fact File: Body Parts 22

Glossary 23

For Further Information 24

Index.. 24

What Are Marine Minibeasts?

Marine minibeasts live in all parts of the sea. A minibeast sounds scary, doesn't it? However, it is really just an animal with no backbone. Sponges, corals, jellyfish, and crabs are among the thousands of marine minibeasts. They come in many wonderful colors and weird shapes. Some are tiny. Others, such as the octopus, can be a few yards (m) long.

Octopus

Sea anemones have soft bodies. The water keeps them afloat.

Many marine minibeasts have a hard shell or **exoskeleton** on the outside. It helps keep them safe from other wild marine animals.

Lobsters and sea stars have an exoskeleton.

Crabs have a hard exoskeleton to keep their soft body safe.

Oysters and clams have a hard shell. When the animal dies, the shell is the only thing left.

Some marine minibeasts just have soft bodies. The water holds up their bodies and lets them float.

Fact Bite

Cuttlefish have their shell on the inside.

Cuttlefish

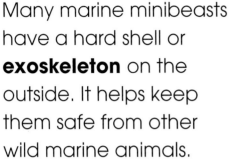

Stuck to One Spot

Barnacles live most of their lives in a shell, stuck to one spot. They make a strong "cement" that glues them to hard surfaces.

They don't hunt for food. They wait for food to float by in the water.

Barnacles mostly live in shallow water. They can also live on the seashore. Some barnacles are covered with water for only a few hours a day.

Fact Bites

Barnacles feed on tiny floating plants and animals called **plankton**.

Barnacles breathe through their legs and body walls.

When barnacles are not covered with water, they keep their shells tightly closed. This stops them from getting dried out by the sun.

The barnacle's shell is made from hard plates that **overlap**. They keep the soft body inside safe. To feed, a barnacle opens a "trap door" in the shell. It reaches out with its weird, feathery legs. They sweep food into the barnacle's mouth.

Each barnacle is both male and female. Wild! While they are young, barnacles swim around. When they are grown, they choose a place to settle. They usually live in large groups.

Barnacles attached to a bottle

Flowers of the Sea

Sea anemones look like wonderful flowers. Really, they are meat-eating marine minibeasts. They live in shallow and deep parts of the ocean.

The sea anemone's soft body is shaped like a tube. At one end is a mouth. There are **tentacles** all around it. At the other end is a base. It can stick to a surface or dig into sand.

These minibeasts are wild hunters. They like to eat small fish, crabs, shrimp, and mussels. When food swims by, the sea anemone grabs it with its tentacles. The tentacles are covered with **stingers** that stun their **prey**. The sea anemone then pulls the prey into its mouth to eat it.

This sea anemone is found in rock pools.

Sea anemones have young in different ways. Sometimes a male and a female have young together. Sometimes a sea anemone makes babies by itself. It splits into two or more pieces. These grow into new sea anemones. Sometimes, a new sea anemone buds from the parent. When it grows big enough, it separates. Weird!

This sea anemone is getting ready to eat a shrimp. It is turning its stomach inside out through its mouth.

Clown fish can live in the tentacles of sea anemones. They are safe here from their enemies. They cover themselves with a special slime so they don't get stung.

Fact Bite

Hermit crabs sometimes "plant" sea anemones on their shell. This protects them from enemies. It's a bit like having a fierce guard dog.

9

Sea Stars

What has arms and feet but no legs, and a mouth but no head? A sea star.

Sea stars are also called starfish. They are easy to spot because they are shaped like a star. Sea stars live on the ocean floor and on **coral reefs**.

The feet of sea stars are full of water.

Fact Bite

Sea stars can be as small as a thumbnail or as big as 3 feet (1 m) across.

These sea stars use their suction cups to stick to the rock.

10

Sea stars move about on hundreds of little tube feet. These feet are full of water. There are suction cups at the end.

Most sea stars have five arms with an eye at each end. These eyes see only light and dark.

Sea stars have a wonderful ability. They can regrow their parts. If it loses an arm, a sea star can just grow another!

A sea star has a weird way of eating. It doesn't put food into its stomach. It pushes its stomach out through its mouth and over its food.

The brittle star is like the sea star in many ways. However, its arms are much longer.

Sea stars have no real head or tail end.

Spiny Sea Dwellers

Sea urchins are shaped like balls. They have big spines. They are part of a group called echinoderms. This means "spiny skin." Sea cucumbers are another type of echinoderm. They are shaped a bit like sausages.

The sea urchin's spines may have **poison** in them. They help guard sea urchins from their enemies. The spines can point in any direction.

The sea urchin's mouth is in the middle of the underside of its body.

This sea urchin moves slowly across a rock. It is looking for food.

Hundreds of tiny tube feet stick out between the spines. These weird feet are full of water. They have suction cups on the end. The feet can only move slowly. The sea urchin is not a fast walker!

Sea urchins are nocturnal. That means they are active at night. During the day, they hide in small holes in rocks or coral. They eat plants and animals. They have strong **jaws** that can scrape **algae** off rocks.

Spines and feet surround the sea urchin's mouth.

This sea cucumber is holding out its tentacles, ready to feed.

13

Sea Slugs

Slugs on land are usually slimy and gray. Sea slugs are often beautiful, with wonderful color patterns. Some of them float upside down on the surface of the ocean.

Some sea slugs eat meat. They feed on sponges, soft corals, sea anemones, and **bluebottles**. Some eat the eggs of other sea snails. Some eat plants.

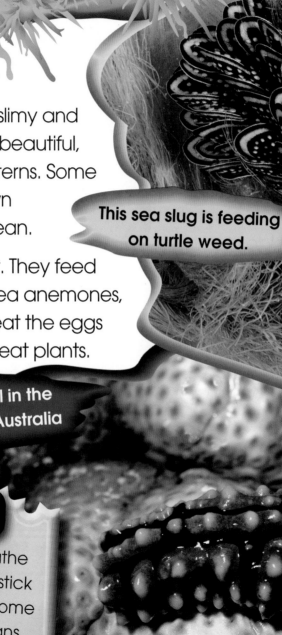

This sea slug is feeding on turtle weed.

A sea slug on coral in the Great Barrier Reef, Australia

Fact Bite

Some sea slugs breathe through **gills**. These stick out on their backs. Some gills look like frills or fans.

Most sea slugs have soft bodies. They have no shell to keep them from being eaten. Instead, their colors and patterns help them blend in with their surroundings. This is called **camouflage**. It makes it hard for enemies to see them. Other sea slugs have very bright colors. This sends the message "I'm dangerous. Stay away!"

Sea slugs have antennae that can smell their food.

Some sea slugs give out a weird smell when they are touched. Their enemies don't like that! They leave them alone.

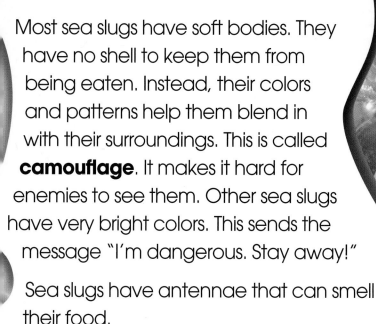

Can you see the sea slug? It is the same color as its surroundings. That's how it hides.

These sea slugs float upside down on the ocean waves. They feed on the tentacles of stingers such as bluebottles.

Sea snails are a lot like garden snails. They have shells to protect their soft bodies. They have weird eyes on stalks. They crawl about on a flat, fleshy foot.

Some sea snails, such as conches, eat plants. Others, such as whelks, eat meat. One type of sea snail is very deadly. It is called the cone shell. Its poison can kill humans.

This sea snail is just coming out of its shell.

Can you see this sea snail's eyes on stalks?

16

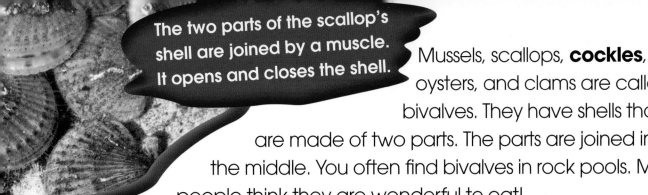

The two parts of the scallop's shell are joined by a muscle. It opens and closes the shell.

Mussels, scallops, **cockles**, oysters, and clams are called bivalves. They have shells that are made of two parts. The parts are joined in the middle. You often find bivalves in rock pools. Many people think they are wonderful to eat!

Scallops have a wild way of moving about. They snap their shells open and closed. Water squirts out. This makes them speed through the water like a jet boat.

Oysters sometimes make pearls. A grain of sand gets caught under the oyster's shell. The oyster coats the sand with a shiny, hard material. This becomes a pearl.

This giant clam has frilly blue lips. It looks a bit like a rock.

The rows of spots you can see in this photo are the scallop's eyes.

Scary Stingers

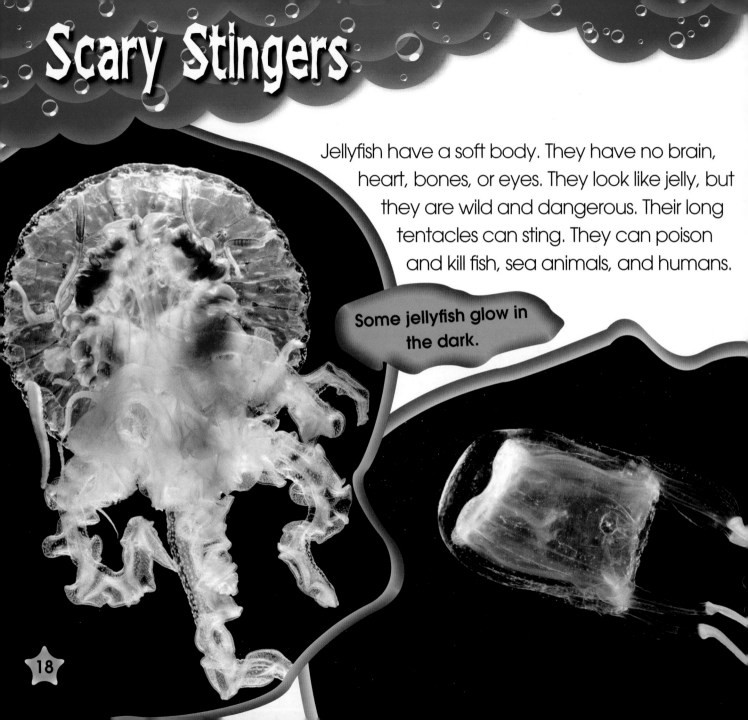

Jellyfish have a soft body. They have no brain, heart, bones, or eyes. They look like jelly, but they are wild and dangerous. Their long tentacles can sting. They can poison and kill fish, sea animals, and humans.

Some jellyfish glow in the dark.

Jellyfish are not really fish. A better name for them is "sea jellies." They are found in every ocean.

Their body is see-through and shaped like a bell. Almost all of the body is water.

Tentacles hang down from the body. Some large jellyfish have tentacles that are almost as long as an Olympic swimming pool.

Jellyfish usually drift along in the water. Sometimes, they open and close their bodies to make the water move around them. This helps bring food close to them.

Box jelly

Jellyfish have long tentacles that sting.

Fact Bites

A group of jellyfish is called a "smack."

One of the deadliest jellies is the box jelly. It kills more people each year than any other sea creature.

19

Floating Hunters

Bluebottles have long tentacles and a float. The float is like a small, blue bag that is full of gas. It lets the bluebottle glide along the surface of the ocean. Wind blows the float, just like a sail on a boat.

Bluebottles are wild hunters. They use their tentacles to catch small fish and sea animals for food. The tentacles have hooks as well as poison.

After a bluebottle catches a fish, it pulls back its stinging tentacles. Feeding tentacles take over.

Fact Bite

Did you know that a bluebottle is not one animal? It is actually made up of lots of animals that live together. Weird!

The snail is feeding on a bluebottle's tentacles.

This floating sea slug is eating a bluebottle.

Bluebottles can give you a nasty sting. They still sting when they are dead and washed up on the beach.

The purple bubble-raft snail hunts bluebottles for food. It has a raft made of bubbles. It is not worried by the bluebottle's stinging tentacles. It thinks they make a tasty treat.

The floating sea slug also hunts bluebottles. It can swallow the bluebottle's stingers without being hurt.

21

Fact File: Body Parts

Marine minibeasts live in the sea and have no backbone. There are thousands of types. They can look very different. Not all marine minibeasts have the same body parts.

Body Parts						
Marine minibeast	**Exoskeleton**	**Shell**	**Soft body**	**Spines**	**Tentacles**	**Tube feet**
barnacle		✓	✓			
sea anemone			✓		✓	
sea urchin	✓			✓		✓
sea star	✓			✓		✓
sea slug			✓		✓	
bluebottle			✓		✓	

Glossary

algae   simple plantlike things that grow in water

antennae   long, thin body parts that stick out from the head

bluebottles   floating colonies of sea animals that work together as if they were one animal

camouflage   using colors, patterns, or shape to blend in to the surroundings

cockles   bivalves that have shells with raised ribs

coral reefs   lines of hard rocks, formed by coral

exoskeleton   a hard covering that supports an animal

gills   the body parts through which fish and other sea creatures breathe

jaws   body parts around the mouth

marine   belonging to the sea

overlap   when something covers part of another thing

plankton   tiny plants and animals that live in the water

poison   something that can kill you or make you sick if you eat, drink, or touch it

prey   an animal that is hunted by another animal for food

stingers   sharp parts of an animal, insect, or plant that go through skin and leave behind poison

suction cups   cup-shaped parts that stick to a surface when pressed against it

tentacles   long, thin, armlike body parts

For Further Information

Books

Aronnax, Pierre. *The Oceanology Handbook: A Course for Underwater Explorers*. Cambridge, MA: Candlewick Press, 2010.

MacQuitty, Miranda. *Ocean*. New York: DK Publishing, 2004.

Web Sites

MarineBio: Marine Life
http://marinebio.org/Oceans/Creatures.asp

PBS: Secrets of the Ocean Realm
http://www.pbs.org/oceanrealm/

Publisher's note to educators and parents: Our editors have carefully reviewed these Web sites to ensure that they are suitable for students. Many Web sites change frequently, however, and we cannot guarantee that a site's future contents will continue to meet our high standards of quality and educational value. Be advised that students should be closely supervised whenever they access the Internet.

Index

barnacles 6, 7, 22

bivalves 17

bluebottles 14, 15, 20, 21, 22

breathe 6, 14

camouflage 15

clams 5, 17

colors 4, 14, 15

crabs 4, 5, 8, 9

enemies 9, 12, 15

exoskeleton 5, 22

eyes 11, 16, 17, 18

food 6, 7, 8, 11, 12, 15, 19, 20, 21

gills 14

jellyfish 4, 18, 19

legs 6, 7, 10

mouth 7, 8, 9, 10, 11, 12, 13

oysters 5, 17

poison 12, 16, 18, 20

sea anemones 4, 8, 9, 14, 22

sea slugs 14, 15, 21, 22

sea snails 14, 16, 17, 21

sea stars 5, 10, 11, 22

sea urchins 12, 13, 22

shells 5, 6, 7, 9, 15, 16, 17, 22

soft bodies 4, 5, 7, 8, 15, 16, 18, 22

spines 12, 13, 22

sponges 4, 14

stingers 8, 15, 21

tentacles 8, 9, 13, 15, 18, 19, 20, 21, 22

tube feet 11, 13, 22